No part of this publication may be reproduced in whole or in part, or stored in a retrieval system, or transmitted in any form or by any means, electronic, mechanical, photocopying, recording, or otherwise, without permission of the publisher.

For information regarding permission:
AppleTree Institute
415 Michigan Ave., NE
Washington, DC 20017

Text and Illustrations copyright © 2015 by AppleTree Institute. All rights reserved. Published by AppleTree Institute.

Printed and bound in Canada.

ISBN 978-1-940641-15-7

Katie Sivinski grew up in Orlando, Florida, and currently resides in the Washington, DC area with her husband, two children, one newt, and many fish. Katie previously worked as an elementary school teacher but found her passion teaching preschool aged children. She now enjoys nurturing her own children and sharing their delight in learning about the world around them. Katie's favorite picture book growing up was *The Little House* by Virginia Lee Burton because she enjoyed seeing the house's feelings and facial expressions progress throughout the book.

Roger James is a trained visual artist who has worked as a graphic designer, illustrator, and oil painter. Roger's formal education includes a degree in art from the University of Maryland, with continued studies in media arts and animation. He has illustrated several children's books, including *Lynn Needs Rest*, *Max Visits the Dinosaurs*, *ABC's of Black History*, and *ABC's of Black Inventors*. Growing up, Roger loved getting his *Berenstain Bears* books in the mail!

Max Visits the Dinosaurs

Written by Katie Sivinski
Illustrated by Roger James

Max loved dinosaurs. He knew all about dinosaurs—names of dinosaurs, what they ate, what they looked like, and even what they sounded like. "ROAR!" Max always yelled.

That's why, one day at school, he couldn't believe his ears. Mr. Meriwether told the class that they were going on a field trip to see DINOSAURS at the Museum of Natural History!

Max was so excited to see the dinosaurs! Mr. Meriwether began to explain that dinosaurs were "extinct," but Max just couldn't listen. He looked out the window and daydreamed…

Maybe on the field trip, they would see a herd of Apatosauruses thundering by, swinging their long necks to reach leaves and plants. Max had learned they were so big they had to eat all day, only resting during short naps.

Or maybe they'd see an Oviraptor. Max knew they were one of the only omnivorous dinosaurs, and they liked to eat clams, lizards, and plants. They were covered with feathers like birds.

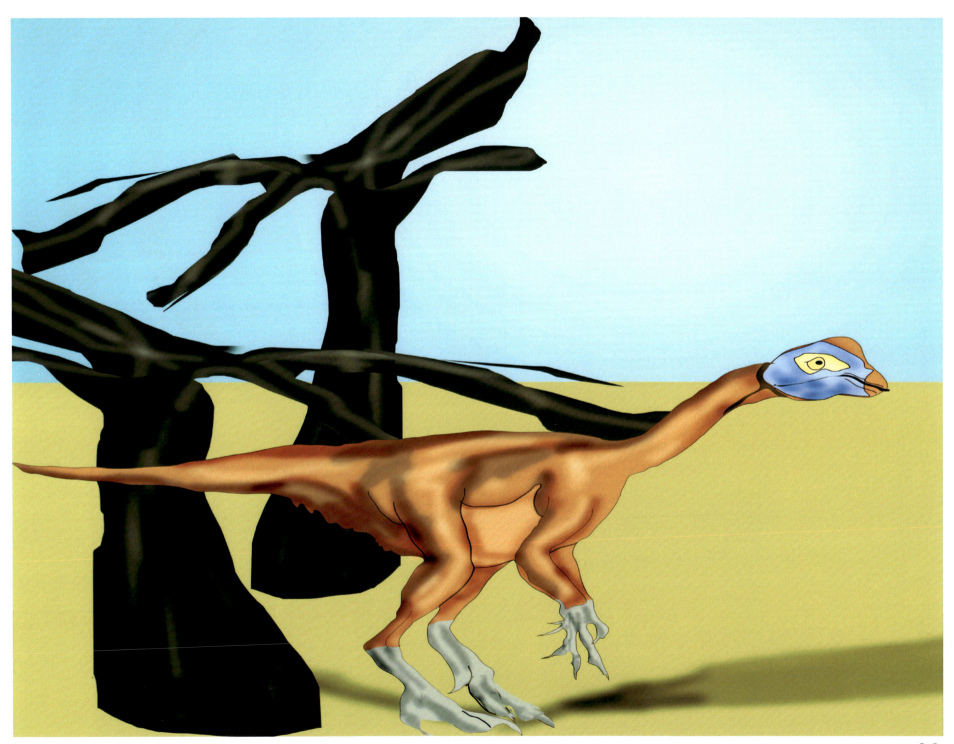

Max wanted to see a Deinonychus because they could run and maneuver fast. He also found it fascinating that they had one large claw on each hand that could retract like a cat's.

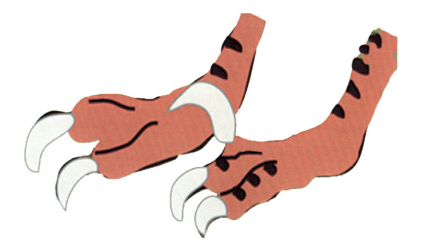

What if they saw the most terrifying dinosaur of all—the Tyrannosaurus rex? Max's dad told him that the Tyrannosaurs rex was one of the biggest dinosaurs of all time. Its body was as long as a tennis court! He also explained that Tyrannosaurus was carnivorous. To catch its dinner, it hid quietly and pounced on unsuspecting dinosaurs.

Max couldn't believe his teacher was taking them on this field trip! All week, he couldn't stop thinking about the dinosaurs they would see. He started to get a little nervous when he thought about their size... and sharp teeth... gulp... giant claws... and brute strength.

The morning of the field trip, Max lathered up with lots of soap in the shower. He even sprayed some of his mother's perfume to mask his scent. He didn't want the dinosaurs to be able to hunt him down with their sense of smell.

As the class got off the bus, Max stayed near the back, hiding behind some of the other kids. He made sure to walk very quietly so the dinosaurs wouldn't hear him.

Once in the museum, he hid behind tall pillars, making his way stealthily so none of the dinosaurs could see him.

Mr. Meriwether spotted him and gave him a worried look. "Max, I thought you loved dinosaurs! What's wrong?" he asked.

"Well, Mr. Meriwether," said Max nervously, "a Tyrannosaurus could come up and eat me. It could eat all of us, it's so big!"

Mr. Meriwether laughed. "No, Max, I told you—dinosaurs are extinct!"

"Extinct?" repeated Max. "What does that mean?"

"Extinct means that all the dinosaurs died millions of years ago. All we have left are fossils—dinosaur bones! The only things people can see of dinosaurs now are models created from these fossils."

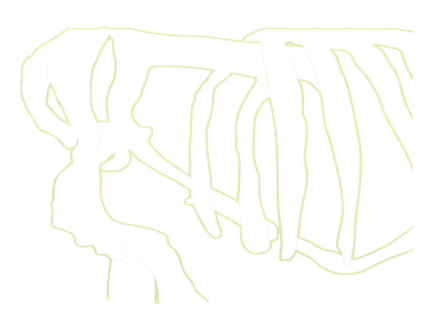

He looked around. He didn't hear any dinosaur roars. He couldn't feel the shaking of the ground as they lumbered around. Maybe Mr. Meriwether was right.

"So the dinosaurs here aren't alive?" Max asked, just to be sure he was safe.

"That's what I mean," said Mr. Meriwether. "There are no dinosaurs alive anywhere in the world. All that's left are their fossils for paleontologists to study."

Max was relieved. He could still enjoy the museum after all. He was a little disappointed to only be seeing dinosaur bones, but at least he knew he wouldn't be eaten. So, Max happily explored the museum the rest of the day.